Our World of Water

WATER
LIFES PRECIOUS WATER
WERE LUCKY
LOOK WHAT OTHERS HAVE
One Planet But Worlds Apart
A World Without Water
Clean Water
SAFE WATER
SAFE LIVES
WATER

Our World of Water

By Joan Stephenson

With a Foreword by
Jacqueline Wilson

1236 Great conduit of Westcheap built to supply London with water.

1581 Peter Morice, a Dutch Water Engineer, developed the first water wheel pump that pumped water from the River Thames.

1591 Sir Francis Drake built 'Drake's Leat', a watercourse that carried water to a conduit in Plymouth.

1600s After being allowed to decline for hundreds of years after the Romans left Britain, piped water supply was re-introduced on a large scale. Pipes were usually made of wood.

1776 James Watt improved the cylinder and piston steam engine that was first built by Thomas Newcommen in 1712. Watt's engine was six times more powerful. This led to the development of steam-powered pumps and greatly assisted in the progress of the industrial revolution.

1829 Sand filtration introduced in England by James Simpson.

1830-1890 Outbreaks of cholera and typhoid in Britain killed thousands.

1848-1875 Public Health Acts were passed to protect public health.

1854-1855 Medical evidence linked cholera and contaminated water. Routine bacteriological testing of water started in London.

Contents

1870 Thomas Crapper designed the flushing toilet – the same basic design that we use today.

1876 Dr Robert Koch proved that bacteria could cause disease.

1897 The first recorded use of chlorine as a disinfectant

1937 All major water undertakers adopted chlorine as a disinfectant.

1974 All the water undertakers were formed into 10 water authorities by the Water Act of 1973.

1981 The British Water Industry created a special charity, WATERAID, to help solve some of the problems of water supply and sanitation in the developing world.

1989 The water authorities were privatized and were now called water companies.

1990 The Drinking Water Inspectorate was formed to monitor the water companies.

2003 South Staffordshire Water plc celebrates their 150th Anniversary.

At the beginning of the 21st Century, the UK population continues to enjoy continuous safe drinking water.

Published by IWA Publishing, Alliance House, 12 Caxton Street, London SW1H 0QS, UK
Telephone: +44 (0) 20 7654 5500; Fax: +44 (0) 20 7654 5555; Email: publications@iwap.co.uk
Web: www.iwapublishing.com

First published 2003

Design and layout by Cato Street Studio, London, UK
Printed by JW Arrowsmith Ltd., Bristol, UK

ISBN: 1 84339 050 7

Foreword by Jacqueline Wilson

I write stories for children. It's much easier writing fiction than fact. You can make up anything you like and as long as you make it seem real then everyone will believe your story. It's just like playing an elaborate pretend game. You can twist things around and always make sure you have a happy ending.

It's much harder writing anything factual. You can't play around with the truth. You have to impart information in an attractive straightforward manner. You have to do careful research and then sift though masses of information, trying to get it all in order. You must make sure something really important doesn't get left out.

I don't think I'd be much good writing a factual account of anything. I'm too used to letting my imagination run riot. I like playing around with language and being inventive. I'm therefore tremendously impressed with all these winning entries for the Drinking Water Inspectorate Children's Competition. I think everyone has tried especially hard.

All the children seem particularly aware that it's vitally important to have a continuous supply of clean water. We're in a privileged position in this country. If we're thirsty we just turn on a tap and gulp down a glass of water without thinking. We take it for granted that we can have a bath and use a flushing lavatory. But there are many people all over the world who are not so lucky. It's therefore very appropriate that a percentage of the profits from this book will go to WaterAid to help provide safe water and sanitation in many communities abroad.

I've enjoyed reading all the competition entries and admired the illustrations, but now I also know a great deal more about drinking water and where it comes from. I've learnt in as easy and entertaining a way as possible – so I'd like to thank each and every one of the competition winners!

Jacqueline Wilson

Acknowledgements

Grateful thanks to Jacqueline Wilson for her invaluable advice and comments. Special mention must be made of Mary Ellis for her extremely helpful suggestions and wonderful way with words. Thanks also go to those other people whose suggestions have helped in the preparation of this book. Last, but not least, a really big thank you to my young friend Holly Ahom who assisted enormously with research, and who gave up many of her weekends to do so.

Joan Stephenson

The Water Companies

In England and Wales there are 26 water companies that look after our water and make sure it is clean and safe. As well as this main job, they also produce a lot of interesting booklets, leaflets, and other material that can be used in schools. Most of the companies have educational websites, which are fun to visit. Some companies invite children to go canoeing, take part in picnics, and even be filmed playing the role of a 'weather person'. There are also 'right to read schemes' where employees volunteer to help local children with their reading. Why don't you find out more about your local water company? Have a look on their website and see what exciting things they have to offer. You can find their website addresses or telephone numbers below.

Anglian Water
www.anglianwater.co.uk

Bournemouth &
West Hampshire Water plc.
www.bwhwater.co.uk

Bristol Water Holding plc
www.bristolwater.co.uk

Cambridge Water plc
www.cambridge-water.co.uk

Cholderton & DistrictWater Company
01980 629203

Dee Valley Water plc
01978 846946

Dŵr Cymru
www.dwrcymru.com

East Surrey Holdings plc
www.waterplc.com

Essex and Suffolk
www.eswater.co.uk

Folkestone & Dover Water Services Ltd
www.fds.co.uk

Mid Kent Water
www.midkentwater.co.uk

Northumbrian Water
www.nwl.co.uk

Portsmouth Water
www.portsmouthwater.co.uk

Severn Trent
www.stwater.co.uk

South East Water
www.southeastwater.co.uk

South Staffordshire
www.south-staffs-water.co.uk

South West Water
www.swwater.co.uk

Southern Water
www.southernwater.co.uk

Tendring Hundring Water Services Ltd
www.thws.co.uk

Thames Water
www.thames-water.com

Three Valleys
www.3valleys.co.uk

United Utilities
www.unitedutilities.com

Yorkshire Water
www.yorkshirewater.com

Wessex Water
www.wessexwater.co.uk

Introduction

Edward Clarke, aged 11 (Anglian Water).

When the Drinking Water Inspectorate, together with 19 water companies in England and Wales, decided to run a competition for school children, the idea was to publish all the winning entries in an informative book. The children were asked to send in drawings, paintings, poems and stories about water. There were a lot of chances to win as prizes were given to winners at schools in each water company area. The winners' schools or classes also got prizes.

As the entries came in, I was amazed and pleased to see how varied they were. All the winning entries are in this book, which also tells the story of water. The original idea has grown and I hope you will find the result interesting, fun and full of information.

In this book you will learn about the scientific origins of water and how the water cycle works. You will also find out about some of the uses that have been made of water through the ages. The book also takes us on a short journey through the history of water in Britain from the time of the Romans until the 21st Century. Today, in England and Wales, great care is taken to make sure that drinking water is safe and healthy. This was not always so and you will find good examples of how people in the 19th Century polluted their water supplies. As a result of this pollution, thousands of people died from diseases such as cholera and typhoid.

You will also learn more about the water companies in England and Wales. You may be surprised to learn that they offer more than just a clean, healthy water supply.

People in other parts of the world are not as fortunate as we are in having safe drinking water. Many people in the developing world still die from waterborne diseases. The World Health Organization believes that about 2.2 million people die each year from these diseases.

Jason Bostic, aged 11 (Thames Water).

The British Water Industry has set up a charity called WaterAid to help provide clean, healthy water, sanitation and hygiene education to the world's poorest people. WaterAid works by helping local communities set up projects that do not cost a lot of money, and that can be managed by the people in the community. WaterAid relies to a great extent on voluntary support and part of the money earned from the sale of this book will go to help in their valuable work.

All the stories, poems and pictures received from the children who took part in the competition were of a very high standard. The children show a good knowledge and understanding of the problems of water in the developing world. They are all a credit to their schools. I hope that the winners, and other readers, enjoy this book.

CHAPTER 1 The Water Cycle

At first sight water looks very uninteresting. It is a colourless, odourless, transparent liquid that we don't think about very much. We turn a tap and fill a glass with water. What happens after that? We can empty the glass by drinking the water or simply by pouring it away. We can leave it in the glass and eventually it will disappear, leaving the glass empty (we call this evaporation and this will be explained a little later).

But where does the water that has been drunk, poured away, or evaporated go? Let's go on a journey of discovery. One thing is for sure, it will turn up again somewhere because water is continually recycled. In other words, it is used over and over again. There are three basic stages in this process, which has the scientific name of the hydrological cycle or, as most people call it, the water cycle.

The three stages of the water cycle are: evaporation, condensation, and precipitation. But first of all, let's take a closer look at what water actually is.

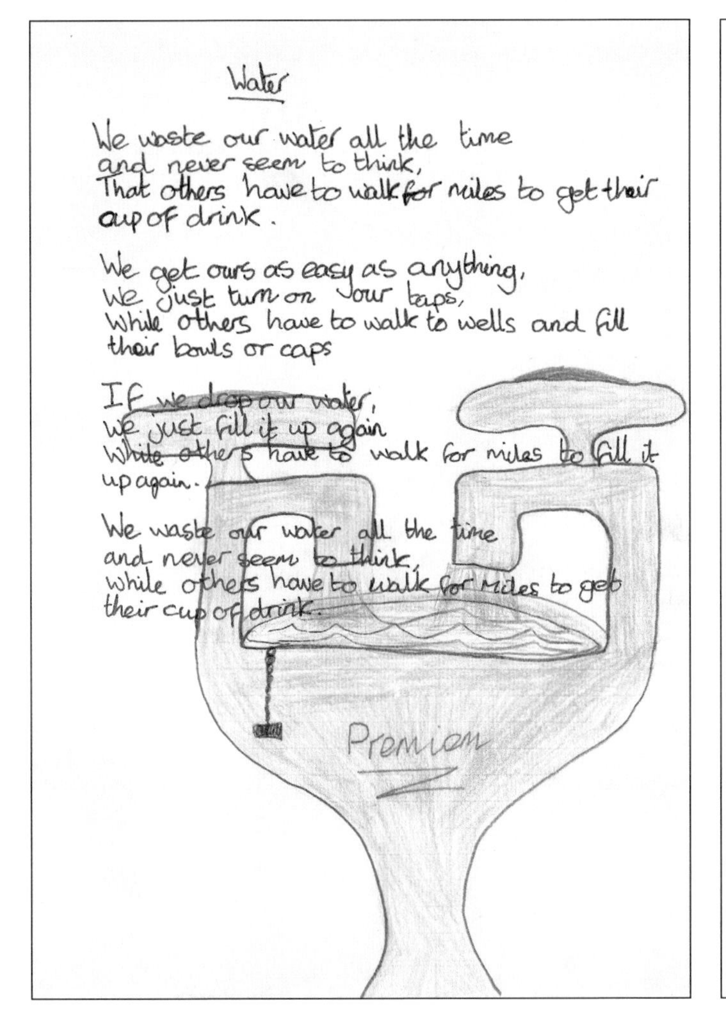

Daniel Nash, aged 10 (Anglian Water).

Katrina O'Connor, aged 9 (Three Valleys Water).

A scientific explanation for the origins of water

Rachael Pennell, aged 11 (Bournemouth and West Hampshire Water).

Scientists have shown that millions of years ago there were no oceans or seas on the planet. The surface of the earth was so hot that any water was in the form of steam, and volcanoes poured huge amounts of steam into the atmosphere. When the earth eventually cooled down millions of years later, the steam turned to water vapour that condensed into droplets and began to fall as rain. The downpour lasted for thousands of years and filled any great hollows that were in the land. These water-filled hollows formed the earth's first seas and oceans.

The nature of water

What is water? Splashy stuff that makes us wet? But is that always the case? Think again. What about ice? What about fog? These are also forms of water.

Water is one of our most precious resources and it is can be found in all three states or forms: solid (ice), liquid (water), and gas (steam). Water's ability to change state when it is heated or cooled is the basis of the water cycle.

What is water made of?

If you were able to look at a drop of water in minute detail you would see that water is made up of hydrogen and oxygen atoms joined together to make water molecules.

Each molecule is made of two atoms of hydrogen and one atom of oxygen, and is written in scientific language as H_2O - the H_2 means there are two atoms of hydrogen and the O means there is one atom of oxygen. H_2O is the chemical formula for water.

These molecules can move about freely, or they can become very lazy and hardly move at all. They can also join together. The ability of the molecules to join together is another important condition of the water cycle.

QUESTION:
Which do you think is lighter, water or ice? Look for the answer in Chapter 7.

Water, Water,
H2O,
Little streams and melted snow.

Water helps us to run and play,
Without it we would not see day,
Little would survive without,
A little water helping out.

Water, Water,
H2O,
Little streams and melted snow.

Water swells with a tidal wave,
Drenching those that care to bathe,
Making them sopping wet,
Although helping those that it has met.

Water, Water,
H2O,
Little streams and melted snow.

Maybe dirty, Maybe clean,
Helping those either kind or mean,
Comes from mountains, Comes from hills,
Often turns in water mills.

Water, Water,
H2O,
Little streams and melted snow.

In can come in by a flood,
Bringing with it tons of mud,
When it goes it leaves a field,
So farmers can improve their yield.

Water, Water,
H2O,
Little streams and melted snow.

You need to drink it to keep strong,
Wash in it to stop a pong,
Without it we soon would die,
Without a chance to say goodbye.

Water, Water,
H2O,
Little streams and melted snow.

Neil Winter, aged 10 (Portsmouth Water Ltd).

Temperature

Water never stays still, it is always moving and changing. Temperature plays a very important role in the water cycle and makes it possible for water to change state.

At low temperatures (below 0 °C) water is a solid. All the water molecules are very inactive and stick together.

Solid

At normal room temperature (around 20 °C) water is a liquid. The molecules are moving about more.

Liquid

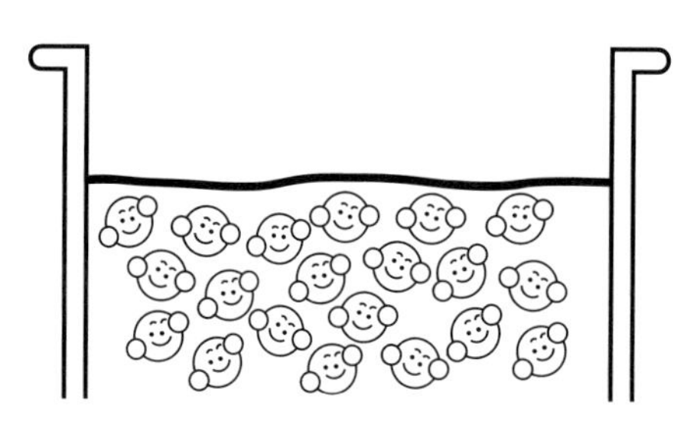

At 100 °C, water starts to boil and is said to have reached its 'Boiling Point.'

At high temperatures (100 °C and above) water molecules move about very freely and change from a liquid into a gas.

Gas

The different stages of the water cycle

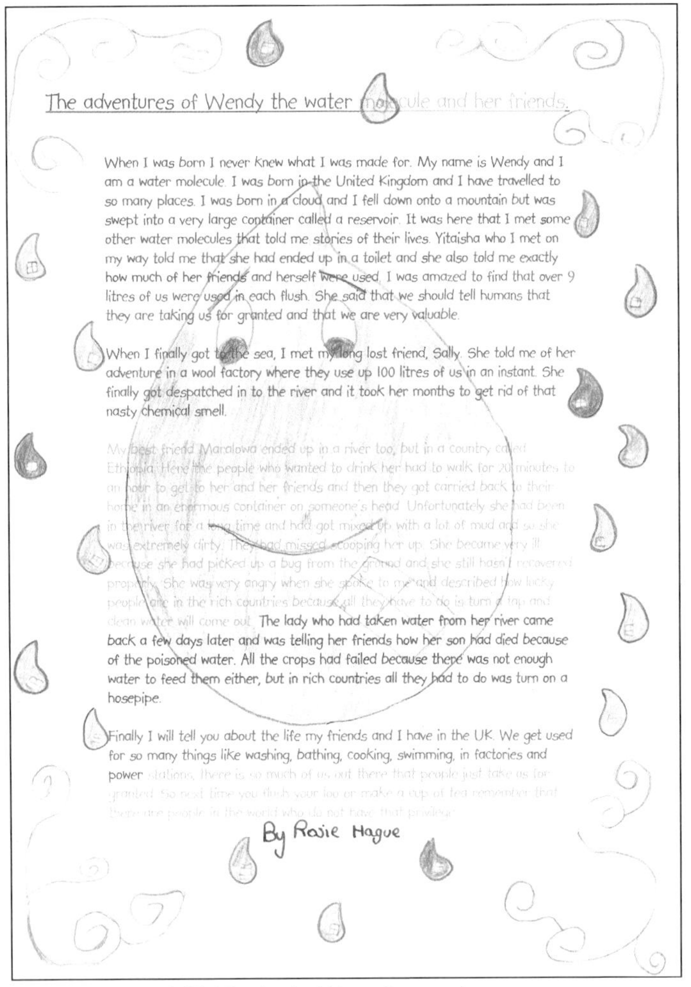

The adventures of Wendy the water molecule and her friends.

When I was born I never knew what I was made for. My name is Wendy and I am a water molecule. I was born in the United Kingdom and I have travelled to so many places. I was born in a cloud and I fell down onto a mountain but was swept into a very large container called a reservoir. It was here that I met some other water molecules that told me stories of their lives. Yitaisha who I met on my way told me that she had ended up in a toilet and she also told me exactly how much of her friends and herself were used. I was amazed to find that over 9 litres of us were used in each flush. She said that we should tell humans that they are taking us for granted and that we are very valuable.

When I finally got to the sea, I met my long lost friend, Sally. She told me of her adventure in a wool factory where they use up 100 litres of us in an instant. She finally got despatched in to the river and it took her months to get rid of that nasty chemical smell.

My best friend Maralowa ended up in a river too, but in a country called Ethiopia. Here the people who wanted to drink her had to walk for 20 minutes to an hour to get to her and her friends and then they got carried back to their home in an enormous container on someone's head. Unfortunately she had been in the river for a long time and had got mixed up with a lot of mud and so she was extremely dirty. They had missed scooping her up. She became very ill because she had picked up a bug from the ground and she still hasn't recovered properly. She was very angry when she spoke to me and described how lucky people are in the rich countries because all they have to do is turn a tap and clean water will come out. The lady who had taken water from her river came back a few days later and was telling her friends how her son had died because of the poisoned water. All the crops had failed because there was not enough water to feed them either, but in rich countries all they had to do was turn on a hosepipe.

Finally I will tell you about the life my friends and I have in the UK. We get used for so many things like washing, bathing, cooking, swimming, in factories and power stations, there is so much of us out there that people just take us for granted. So next time you flush your loo or make a cup of tea remember that there are people in the world who do not have that privilege.

By Rosie Hague

Rosie Hague, aged 10 (Cambridge Water Company).

1. Evaporation

Evaporation is the name given to the process where water changes from a liquid to a gas. The sun warms up the water in the world's oceans, seas, lakes and rivers and evaporation takes place quite quickly. Some of the water changes from liquid to an invisible gas called water vapour. The air above the surface of the sea is also warmed by the sun, and it rises into the sky carrying the water vapour with it. Water will also evaporate on a cool day, but at a slower pace.

When there is a lot of water vapour in the air we feel hot and sticky, and the weather is described as being humid.

Water also evaporates from plants. In a large forest an enormous amount of water will evaporate from the leaves of trees and shrubs. This is called transpiration.

Faster evaporation takes place when the weather or climate is hotter or when wind speeds are high.

The sun isn't the only source of heat that makes water evaporate but it is the major cause in the water cycle. Heating a pan of water on the stove will also cause evaporation.

2. Condensation

Condensation is the opposite of evaporation. The higher you go up into the sky the colder the air becomes. That is one of the reasons that you get snow on top of mountains. So, as the warm air containing the water vapour rises higher, the cooler it becomes. Cooler air cannot hold as much water vapour as warm air and the water vapour begins to turn back into tiny droplets of water. This is known as condensation. (Think of how vapour from a boiling kettle condenses when it hits a cooler surface such as a window pane.)

Clouds are formed of water vapour that has condensed. In other words, clouds are made up of billions of tiny droplets crowded together. In very cold temperatures, these droplets of water may become ice or snow.

QUESTION:
What sort of temperature do you think is required for greater evaporation to take place?
Look for the answer in Chapter 7.

3. Precipitation

Precipitation is the name given to the part of the cycle where water falls back to earth in its various forms as rain, sleet, snow or hail. Precipitation occurs when the water droplets in the clouds join together to form larger and larger drops. As the drops get larger they become heavier. The wind blows the clouds along, and they in turn join together. Eventually the drops become so heavy that they can no longer remain floating up in the sky as the pull of the earth's gravity is too strong. It causes the drops of water to fall down to earth.

Some of the water that falls to earth will stay on the surface and run into streams, rivers, lakes, reservoirs, and seas. This type of water is called surface water.

Quite a lot of rainfall will soak into rocks and soil. This water is stored in rocks and is called groundwater. These rocks are like massive sponges, with tiny holes known as pores, and are known as aquifers.

Some of the water which falls as snow may remain frozen and become part of an ice sheet or glacier at the North or South Poles. The deeper layers of ice in some glaciers can be over 100,000 years old.

Eli Jenkinson, aged 10 (Southwest Water).

QUESTION:
What are clouds made of? Look for the answer in Chapter 7.

Jamille Smith, aged 9
(Dŵr Cymru Welsh Water).

Water Is Life

Clean water is what *we've* got
Other countries, they have not.
They've got water full of mud
I'll carry on explaining until you've understood.

They have to travel miles each day
Sometimes they even have to stay
in a long, long queue.
You wouldn't like that to happen to you!

Such hot countries – they're entitled to a sip
Some don't even get a drip.
Dirty water spreads infection
Can't be cured with an injection.

They surely shouldn't have to fight
Clean water is everybody's right.
But what we do
is up to you!

Stop all the pollution
that's one solution.
Greed is what we don't need
Let's be fair –
Give them their share.

Now it's time
to end my rhyme
This is what I want to say
I want you to think
Every time you drink …

"Water is life …
Don't let it drip away"

Ryan Al-Hakim, aged 9 (United Utilities).

QUESTION:
Does the water cycle ever stop?
What do you think would happen if it did stop?
Look for the answer in Chapter 7.

Water Poem

Just think how lucky we are!

We can go to the kitchen and have a drink.

We can stand under the shower and soak.

We can get up every morning and have a wash.

Just think how lucky we are!

We can put on clean clothes in the morning.

We can wash up our dirty dishes.

We can go to the toilet when we need to and flush.

Just think how lucky we are!

We can cook delicious food.

We can swim in the swimming pool.

We can freeze our food so it stays fresh.

Just think how lucky we are!

BUT

We can be very self-centred.

We can be very wasteful.

We can be very careless.

Just think how selfish we are!

We can be very uncaring.

We pollute our rivers and streams.

We waste our food and drink.

Just think how selfish we are!

Just think how hard it must be!

They have to walk for miles and miles.

They have to drink dirty water.

They have to carry big buckets.

Just think how hard it must be!

They have to wait 15 minutes to drink.

They starve from hunger and thirst.

They cry out for water.

Just think how hard it must be!

By Orla Cronin

Orla Cronin, aged 9 (Tendring Hundred).

Types of water

There are two types of water found on earth: salt water and fresh water. Over 97% of the water on earth is salty and is found in the oceans.

A huge range of marine plants and animals live in the oceans. They are well-adapted to living there. However, fresh water, which does not contain salt, is essential for life on the land. Less than 3% of water is fresh water.

Hard or Soft?

As well as being salty or fresh, water can be hard or soft depending on the area you live in.

Hard water

Some substances dissolve quite easily in water. If, after rainfall, water passes through soft rocks like chalk and limestone, it will pick up minerals such as calcium and magnesium that are found in these types of rocks. The minerals are good for the development of teeth and bones, but make water 'hard'. It does not mean that the water will feel hard to the touch. It means that it will be harder to get a good lather from soap. Hard water leaves scale on the elements of kettles because when

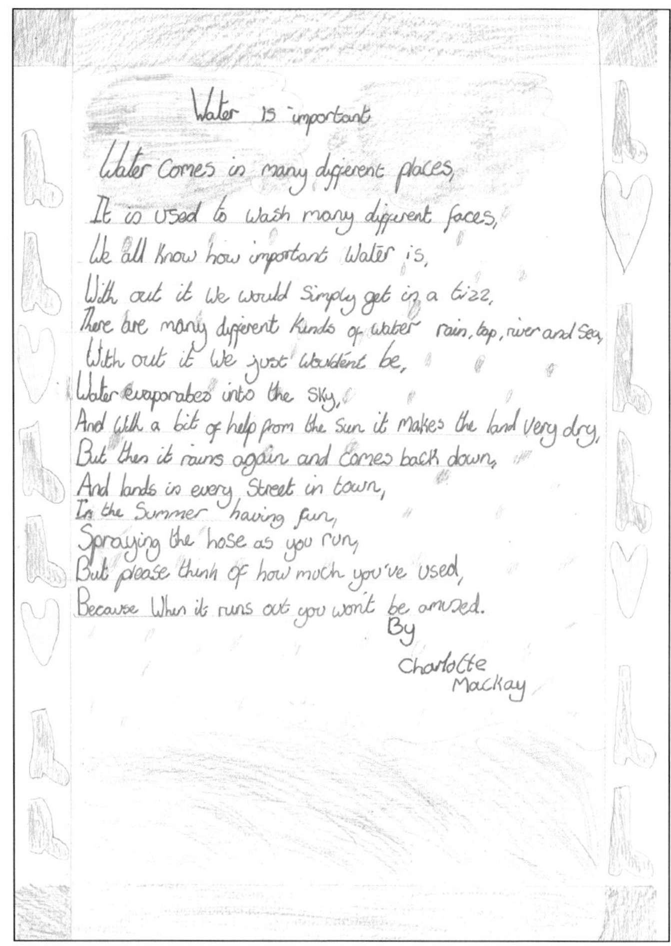

Charlotte Mackay, aged 10 (South East Water).

Esther Dobson, aged 10 (Thames Water).

the water is boiled it evaporates and leaves the minerals behind. Have a look at the inside of an empty kettle when it has cooled down. Can you see any limescale?

Soft water

Soft water is water that has passed through hard rock such as granite. It will not pick up any minerals because the rock is too hard. It is easier to use with soap and makes a good lather, unlike hard water.

QUESTION:
Can you name the world's oceans?
Look for the answer in Chapter 7.

Did you know?

- Small rain drops are spherical.
- Larger raindrops assume a shape more like that of a hamburger bun.

Availability of water

We have already seen that over 97% of the world's water is salty. The remainder, nearly 3%, is freshwater: almost 2% is stored as ice at the North and South Poles and in ice sheets; that leaves about 1% for our use.

Although it seems that very little fresh water is available for plants and animals, there should be enough to meet all our needs. Unfortunately, fresh water is not evenly spread across the world. Some places are very wet, some are very dry.

The amount of water available for people depends on:

- where they live
- the time of year
- the number of people living there.

In countries where rain falls regularly and temperatures are relatively low, water supplies are generally good. Northern Europe, for example, has plenty of rain.

Light rainfall and high temperatures produce dry conditions and water shortages. When the rain falls it evaporates away very quickly because of the high temperatures. There is often little vegetation to trap the water and the soil may be dusty and unable to hold the water that does fall. Large parts of Africa and Asia, for example, have these conditions and life can be difficult, particularly when there is a drought.

QUESTION:
We have talked about droughts happening when there is not enough rainfall. What do you think would cause a flood?
Look for the answer in Chapter 7.

The world's water

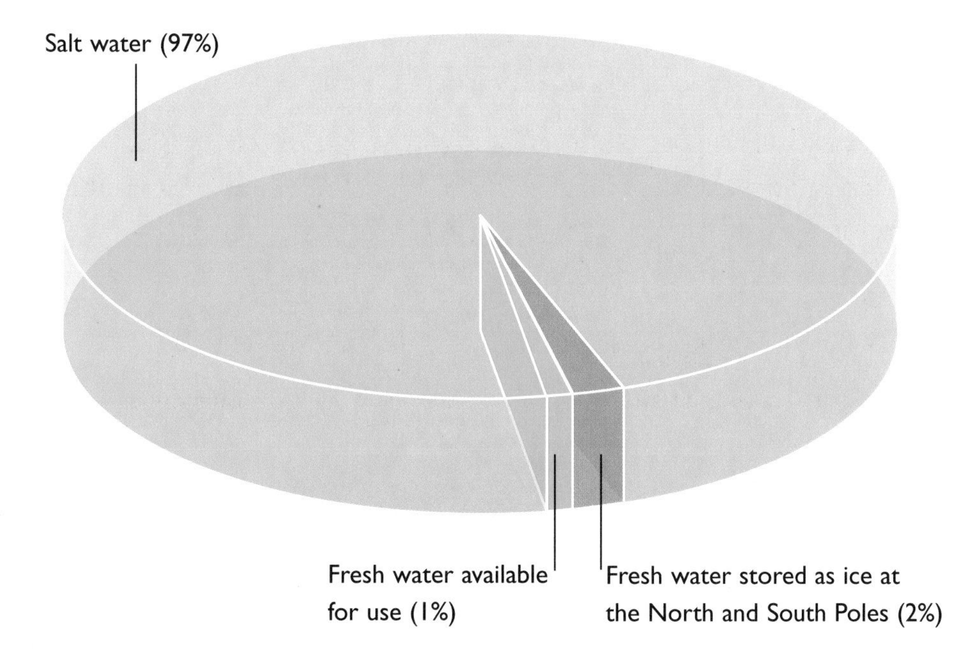

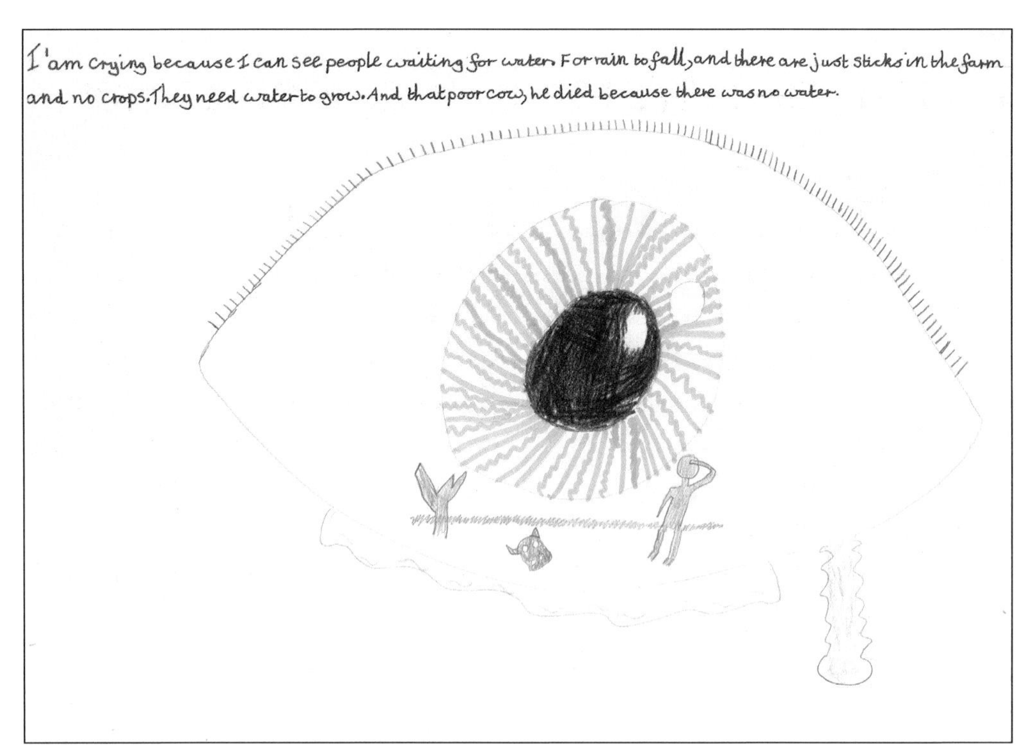

Abbie Jones, aged 9 (South Staffordshire).

Dear Mum and Dad Emma Allsop

I have been out here for months and the weather is getter hotter and drier. It has not rained here for 8 months and the crops have all died off in the fields.

The work on our project has gone well but it is hard work in this heat. Compare our use of water to theirs and the difference is amazing. I used to get up, back in England, leave the tap running whilst cleaning my teeth and having a wash. Long hot showers, helping dad wash the car weekly and using a hosepipe to keep the grass green and the flowers growing. Over here water is treated like "gold", it is so precious it is never wasted. They pray daily for it to rain, even have a god of rain. Water has to be collected by buckets from the local river. This is a little walk of 2 miles there and 2 miles back and it is the children to go to fetch the water. They go upto 3 times a day. This water collected is not clean and can make people ill. As you read this letter and Mum goes to put the kettle on with fresh clean water from the tap, think of these poor people.

However, all that is to change.
Our well is finished and tomorrow using solar power to run the pump, fresh water will be available for all the village to use. The water treatment kits will make it safe to drink, and by the end of the month ditches will have been dug, allowing water to be pumped to the fields so that new seeds can be planted and will grow.

I must go now as more digging is to be done, having a great time and remember how lucky we all are back in England to have lots of water.
Lots of love

Emma

xxxxxx

Emma Allsop, aged 11 (Severn Trent Water).

CHAPTER 2 Water in Ancient Times

People use water for many different things. It is very versatile. Here are just a few examples of what water can be used for: drinking, bathing, farming, cooking and transportation.

Water can absorb a lot of heat before it begins to boil and that makes it very useful to industries. It is used to cool down all sorts of things such as hot steel when it comes out of a furnace. It also keeps our car engines cool. Water is also a great solvent – which means that lots of other substances can be dissolved in it, such as sugar or salt.

These are some of the uses of water. Can you think of other things that it can be used for?

Sarah Whitehead, aged 10 (Southern Water).

Water collection and use in the past

Let's go back in time and have a look at how people collected and used water centuries ago.

Pre-history Humans drank straight from rivers and lakes, just as animals do today. They soon learned to collect water in large jars or other water-tight vessels and carry it to where it was needed.

Some uses for water

Drinking
Bathing
Farming
Cooking
Transportation

4000 BC The country that we now know as Iraq was known as Mesopotamia. Mesopotamia was divided up into several areas, each area having its own King. The people living in these areas built dams for irrigation and canals were dug to transport water for domestic use and for farming.

2500 BC People living in the Indus Valley, the area that is now Pakistan and North West India, had a very advanced plumbing system. They had brick pipes and sewers and an elaborate drainage system under the cities. They stored their water in brick-lined wells.

1760 BC In Babylon, one of the cities of Mesopotamia, the King passed laws to make sure that canals for transporting water were built all over his city. His laws also made sure that the canals were kept clean and not neglected. He appointed officials to enforce the law.

1000 BC The Egyptians invented a device for raising water from one level to the next. Hundreds of years later a Greek, called Archimedes, improved this device and gave it his name: 'Archimedes screw'. It consists of a giant screw inside a cylinder. One end of the cylinder is placed in the water and the other end is placed at the higher level. When the screw is turned, water is drawn up. This method is still being used today.

705 - 681 BC Another area of Mesopotamia was called Assyria. The Assyrians were very warlike and eventually conquered nearly all of Mesopotamia and built up a huge empire. The first aqueducts were built here. An aqueduct is a bridge with water channels running along the top. Aqueducts are usually built to cross valleys, taking water from rivers or streams to cities where the water is needed.

600 - 565 BC A king of Babylon loved his wife so much that he decided to build a huge garden for her. The king's name was Nebuchadnezzar - see if you can pronounce it! This garden was built on a steep slope on different levels. The king's engineers devised irrigation machinery that took water to the top of the garden and then allowed the water to run down the levels and keep the soil moist. There was a drainage system to take away excess water. The gardens became famous and are known as the 'Hanging Gardens of Babylon'. They no longer exist and people are still searching for the exact site where they might have been.

Claire Leeson, aged 11 (Yorkshire Water).

Never take things for granted

How would you feel with no water?
Would you feel sad or angry or not?
Always be contented.
With the water you have got.

People overseas are suffering,
They have no water left,
It has all disappeared,
Just like a giant theft.

But we don't know what its like,
To have no water I mean,
As everywhere we go today,
There's water to be seen.

They don't complain about walking,
Around ten miles a day,
Carrying water in a pot,
A pot made out of clay.

Africa, Asia and Kenya,
Are some of the countries like this,
If someone asked if you'd like to go,
You'd say I'll give it a miss.

So next time you drink water,
From the faucet or the tap,
Just think what it would be like,
If it never came back.

By Claire Leeson

It might be fun trying to find out more about the 'Hanging Gardens' and what was special about them. Ask your teacher to tell you about them, or tell you where you can find out more. The internet would also be a good place to start looking.

400 BC It is believed that the Egyptians were the first people to keep records of the methods of cleaning water. They would clean water by boiling it and then filtering the boiled water by passing it through sand and gravel. They also used a substance called alum to remove solids (large particles) from water. Alum is used in our water treatment today to remove solids.

3 BC Public water supply systems were established in Rome, Greece and in Egypt. Water was cleaned by filtering it through coal and pebbles or sand.

100 AD The Romans adopted the Assyrian methods and built spectacular aqueducts. They also recognized the importance of good water supply and good sanitation. Nine aqueducts brought over 1,000 million litres of water to the city of Rome every day. Vast networks of clay sewers were developed to take waste water away. The Romans were also the first people to make pipes from lead.

This is the famous Pont du Gard aqueduct built by the Romans, that carried fresh water across this river valley on its 50 km journey to the city of Nimes in the south of France.

QUESTION:
Do you know what AD means and what language it is from?
Look for the answer in Chapter 7.

Unusual uses of water

There have been some unusual uses of water. Here are two examples:

An Egyptian water clock

Egyptian water clock

The Ancient Egyptians were quite a sophisticated bunch and obsessed with time. They invented several clocks and the water clock was one of them. The water clock was made from two bowls. One bowl had markings on the inside that represented 'hours' and had a small hole in the bottom. This bowl would be filled with water that would drip through the hole into the other bowl that hung beneath. The markings in the bowl would show how much time had passed as the water level got lower. Water clocks are still in use today. There is one at the Bristol Water Company in Bristol.

Maybe you could ask your teacher if the class could build a water clock. *See page 35 for instructions.*

Chinese water torture *Do not try this at home or anywhere else!*
Everybody likes to have a shower. It is fun to have thousands of droplets of water washing you clean, but what about one little droplet, falling on your forehead, for hours and hours and hours?

Many, many years ago people used to use a form of torture called the Chinese Water Torture. The victim was tied up with water dripping on his

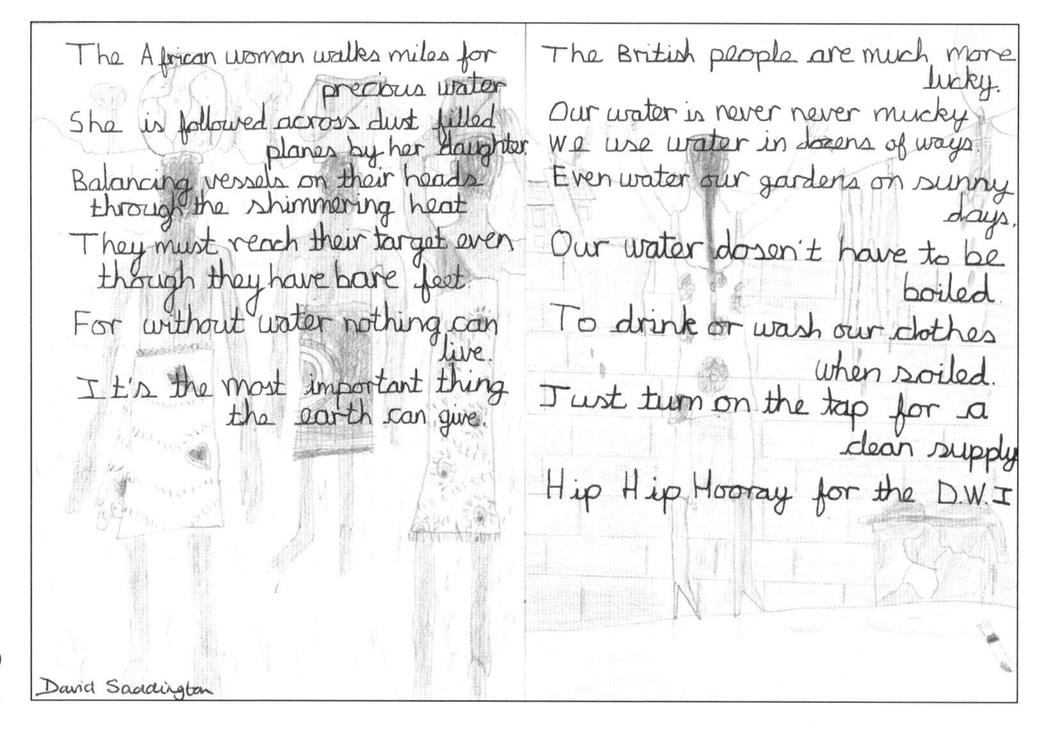

David Saddington, aged 10 (Northumbrian Water).

forehead. It doesn't sound very frightening, does it? At first the victim wouldn't mind. What harm could one little droplet do? But what would happen when he wanted to go to sleep? Drip, drip, drip. What would happen when it made his forehead itch, but he couldn't scratch because his hands were tied. Drip, drip, drip...

This was a very effective way of making a person tell secrets or confess to doing things whether he was guilty or not – and all because of one little drop of water.

8th March
Dear Diary,
I hate my life. I got up at 4:00 am, so I could get ready for breakfast. I was watching the time as I got dressed so I could be at the front of the queue. I am the only one in our family that gets up early to fetch water! I had to walk miles and miles and miles to fetch the water. To make it worse, I was barefoot and I had three buckets with me. I also had to keep stopping. Six times, actully, but I survived! When I finally arrived I queued, queued and queued. It took an hours waiting and when it was my turn to fetch water, there was only half a bucket full left. I didn't even have that much because there was at least twenty more people behind me. It is so not fair that people have to walk miles for hardly anything.
Oh, guess what? I have been punished. I have been badly punished and I am crying my eyes out. Why? I'll tell you right from the start... I was half way through my journey when this boy ran past me at top speed. He was running so fast he tripped me up and all of our water spilled and dried up in the sand. He blamed everything on me and started pushing me around saying how his family won't have anything. He was shouting for so long that it made me really late. That's why I've been punished.
See you tomorrow with another heart breaking story.
Love Karishma x

Beth Rimmer, aged 11 (South Staffordshire).

The Water in the third World

It is disgraceful what their water is like,
They have to walk, they can't go by bike.
There is no water, river or seas near,
Dying from their own water must be a fear.
While the young children are dying,
Their mothers are loudly crying.
They have to walk ten miles to drink,
It doesn't even make you stop and think,
Of what you are doing by throwing it away,
While little children sit and pray.

Levi Scott, aged 10 (Southern Water).

These are just some of the few inventions that have taken place over hundreds of year in the use of water. See if you can find out more about these inventions or others that have to do with water.

Annie Jenner, aged 10
(Thames Water)

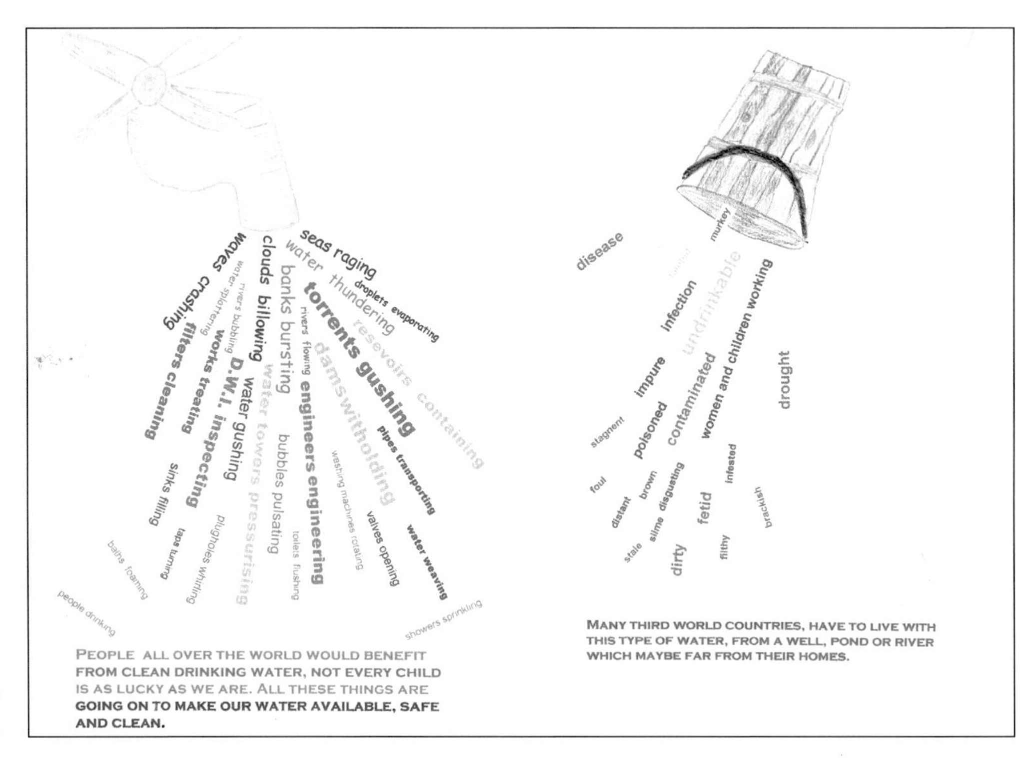

Miles Macallister, aged 10
(Essex and Suffolk Water)

CHAPTER 3 Britain after the Romans

The Romans occupied Britain from about 43 AD to 400 AD. They were a very advanced civilization and took their way of life to all the lands they invaded. When the Romans conquered most of Europe and Britain they brought piped water supplies, drainage and public bathing houses. There is still a Roman bathing house in the town of Bath, today. The Romans dedicated the town to one of their goddesses, Minerva, and called the town Aqua Sulis.

Kyle Goddard, aged 11 (Cambridge Water Company).

They also built aqueducts to bring water from rivers to towns where they settled. They made life very easy for themselves although most of the native Britons continued to collect water from rivers and streams.

The story of Britain's water took a dark turn after the Romans left. Out went all the good work that they had done. Water and sewage systems were abandoned and sewage was thrown out onto the streets or into the streams and rivers. People went back to primitive ways. Very few people bothered to have a bath, especially in winter!

As towns began to develop and expand people found that they had to live further waway from water sources. Water sellers would bring water to the towns and richer people would buy it from them.

However, water supplied by water carriers soon became insufficient as the amount of people living in the towns increased even further. In the 1600s the government appointed water undertakers to lay pipes to bring water to the cities. (They were called water undertakers because they 'undertook' to provide

Dear Mr Blair,

I would like to draw your attention to a problem in many parts of the world, especially in Tanzania – the lack of safe drinking water.

Many people in Tanzania have to walk six kilometres to find water; it is ruining their lives. They should be able to enjoy times with their families but collecting water is taking all their time as it is so important to them. Many of them have to get up early in the morning to go and find water; this should not be happening. They don't really have a choice as water is needed for life.

If charities like WaterAid could build some more hand pumps in villages in Tanzania people would be able to live a safer and happier life. They would be able to spend time with their friends and would not have to walk so many miles each day to find water. It would mean they would be able to keep clean, stay healthier and enjoy safer drinking water.

We need to remember that some people are not as lucky as we are and we should be treating our water the right way. We should not leave the tap on while brushing our teeth and we can use a hippo in the toilet to save water. Having a shower instead of a bath would help too. If we use our water the right way we will not waste so much of it.

We also need to be more careful about not polluting our water. Our water often gets mixed with pesticides, fertilisers and petrol. It takes time and money to get it out of the water. Pollution in seas and rivers kills plants and wildlife. We need to be more careful and treat our large amounts of water properly.

Yours sincerely

Mairead Greene

Mairead Greene, aged 9 (Thames Water).

Chloe Hardy, aged 9 (Yorkshire Water).

water.) The water was usually piped to central fountains (called conduits) where people could collect it. Some richer people had pipes laid to their houses. These pipes were usually made of wood, especially from the Elm trunk. However, there were no sewers. People would throw their household waste into the street with a cry of 'Gardez L'eau' (meaning 'beware the water'). Usually they shouted after the waste was thrown and the unfortunate passers-by were drenched with waste from chamber pots and with kitchen slops!

Water pollution increased alongside the growing populations in cities. People still emptied their waste onto the street and into the rivers. Rivers all over the UK became polluted, especially the River Thames in London. The Thames was very dirty, smelly and full of rubbish. Water from the river was usually called 'monster soup'. The Houses of Parliament are near the Thames and in 1858 the smell from the river was so bad that the politicians had to close Parliament. The terrible smell caused that year to be known as the 'Year of the Great Stink.'

QUESTION:
What caused the water from the Thames River to be called 'monster soup'? Look for the answer in Chapter 7.

The water undertakers would supply drinking water from the same rivers into which waste was being emptied. The water was very dirty and full of bacteria and it was not disinfected or filtered before people drank it. They did not know that diseases could be caught from dirty and contaminated water.

Cholera in 19th Century Britain

Between 1831 and 1866 there were outbreaks of a disease called cholera in which over 100,000 people died. Cholera killed more people than any other disease during the mid-19th Century. It was caused by drinking dirty water that had been polluted by sewage. It had still not been discovered that germs caused diseases. Many people believed that cholera was sent by God as a punishment for wickedness. Others believed that it was caught by inhalation of 'bad' air and people would walk around with a small bunch of flowers (called a nose-gay) held to the nose to keep the dreaded cholera away – this did not work!

Strange as it may seem, brandy was prescribed as a cure and as a preventive medicine for cholera. Of course, the people who sold the brandy were scared to come in contact with the cholera victims buying it. One establishment made a hole in the wall of their shops through which they sold the brandy to the cholera sufferers. Their premises became known as 'The Hole in the Wall'. The sufferers got very drunk on the brandy but still died.

Matthew Boucher, aged 9 (Portsmouth Water Ltd).

A doctor called John Snow was able to show the government of the time that there was a link between the cholera outbreak and contaminated water. While dealing with an outbreak of cholera in London in 1854, he realised that most of the people who caught the disease drank from the water pump in Broad Street in London's East End. He even noted that one lady who caught the disease lived quite far from the pump but had water brought to her from the pump because she liked the taste! She died. John Snow removed the pump handle so the pump could not be used and the number of deaths went down.

All change

Something had to be done, but what? The water supply had to be made clean. The Government came under increasing pressure to do something about the state of the water supply. It passed public acts to clean up water sources so that the health of the population could be protected. In the 1860s a sewerage system was built in London. It is still in use today.

To ensure that water was always plentiful, engineers began to create reservoirs by building dams across valleys. The first reservoir in Europe was built in Longendale in the Pennines between 1848 and 1877. There are now over 50,000 reservoirs around the world.

Zara Kershaw, aged 10 (United Utilities).
Zara says: 'We can get water just from a tap. Some other people are different. This picture shows different points of how we use water and things that need water. For instance the girl in the right-hand corner has to walk miles sometimes just to get a bucket of water to live on.'

To Bathe or not to bathe

Although water supplies were improved, bathing was still not a common practice. Right up to the middle of the 19th Century, some of the poorer people, who usually had only one set of clothes, would have themselves sewn into their clothes at the start of winter to keep warm. They would not wash for the whole winter. Can you imagine the smell? Even some of the rich people would not wash. They would simply change their clothes and use a lot of perfume. The Duke of Wellington, who took a bath every day, was regarded as an eccentric.

QUESTION:
Do you think that we have benefited from inventions made by ancient civilisations? Can you name any of these that are in use today?
Look for the answer in Chapter 7.

Did you know?

- *There were no bathrooms in Buckingham Palace in 1860.*

Did you know?

- *The water closet, which was the name given to the first toilets, was revolutionised in the 1870's when Thomas Crapper and Company designed the flushing toilet system which is still in use today.*

CHAPTER 4 Water in the 20th and 21st Centuries

Drinking Water

Have you ever thought about how clean water comes out of the tap whenever you turn it on?

Luke Turpin, aged 11 (South East Water). Luke drew this picture after visiting a water treatment plant with his school.

Rainwater is not very clean when it is found in nature. It contains substances that have to be removed before the water can be used. It also contains tiny particles that we cannot see with the naked eye. These tiny particles are not fit for humans to drink. The discovery at the end of the 19th Century that bacteria caused waterborne diseases such as typhoid fever and cholera meant that at last scientists and engineers could set about getting rid of these diseases.

Today there are 26 water companies in England and Wales. They play a very active part in the process of the water cycle by providing clean, uncontaminated water to people's homes. The development of technology has enabled these water companies to clean water using a very thorough and highly complex process.

QUESTION:
Water that is stored in rivers and streams is called surface water. What is water that is stored in underground rocks called? Look for the answer in Chapter 7.

Water is taken from its source, either surface or groundwater, and any large particles are removed. Chemicals, such as alum, are added to help remove smaller particles in the water. The water is then filtered through layers of sand

Hannah Wright, aged 10 (Three Valleys Water).

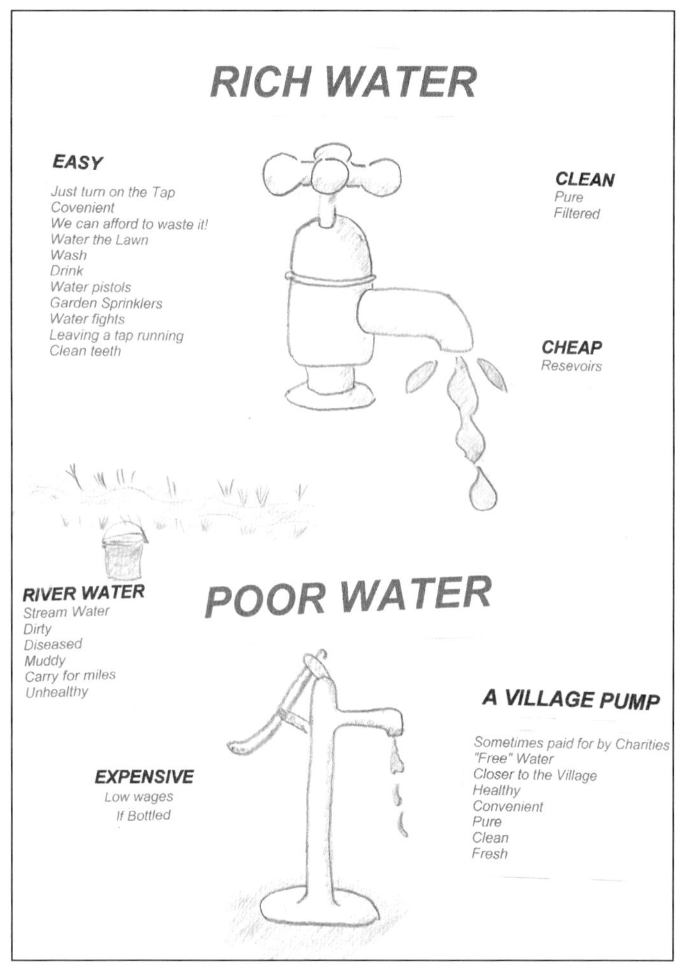

Bethany Eaton, aged 10 (Bristol Water).

QUESTION:
Can you remember what the rocks that store water are called?
Look for the answer in Chapter 7.

and gravel to remove any remaining particles that might still be in the water. The chemicals are also removed after they have done their job. A small amount of chlorine, which is harmless to humans, is added to kill any remaining germs and to keep the water safe as it travels to our homes. Water is checked thoroughly at each step in the process by technically advanced equipment to make sure that there are no harmful substances in the water.

Water is then put into distribution systems, ready for us to use by simply turning on a tap. After we have used the water it is taken away through drains to a waste water treatment plant. Once there, it is cleaned and any harmful substances are removed. The water is returned to the rivers, which then take it back to the sea where it enters the water cycle again.

CHAPTER 5 Water around the world

We are very lucky in the UK. Water companies provide drinking water that is clean and healthy and the government has set up an organisation called the Drinking Water Inspectorate to check that the water companies keep it that way. But over one billion people living on this planet do not have any form of

Harrison Fisher, aged 9 (Bristol Water)

treated water supply. In most developing countries, water still has to be collected from wells, lakes and streams that are often very dirty. The water then has to be carried, sometimes over long distances, back to villages and homes. Some water sources dry up, making the journey for water longer and harder. Every day over 25,000 children in developing countries suffer from illnesses caused by drinking polluted water.

WaterAid, a UK charity set up by the British Water Industry, works with communities to improve water and sanitation. For more information about WaterAid, have a look at their website: www.wateraid.org.uk.

Many of the following pictures, poems and letters in this book are from children in England and Wales who are concerned about the problems of water in the developing world. The entries in this chapter are no exception.

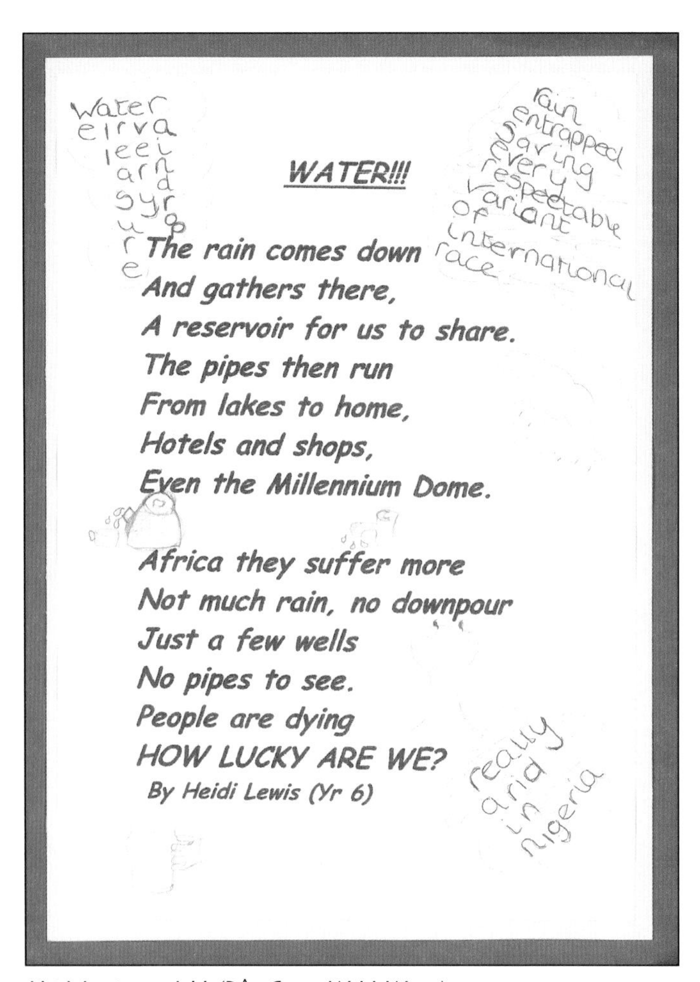

WATER!!!

The rain comes down
And gathers there,
A reservoir for us to share.
The pipes then run
From lakes to home,
Hotels and shops,
Even the Millennium Dome.

Africa they suffer more
Not much rain, no downpour
Just a few wells
No pipes to see.
People are dying
HOW LUCKY ARE WE?
By Heidi Lewis (Yr 6)

Heidi Lewis, aged 11 (Dŵr Cymru Welsh Water).

Sophie Baldwin, aged 10 (Bournemouth and West Hampshire Water).

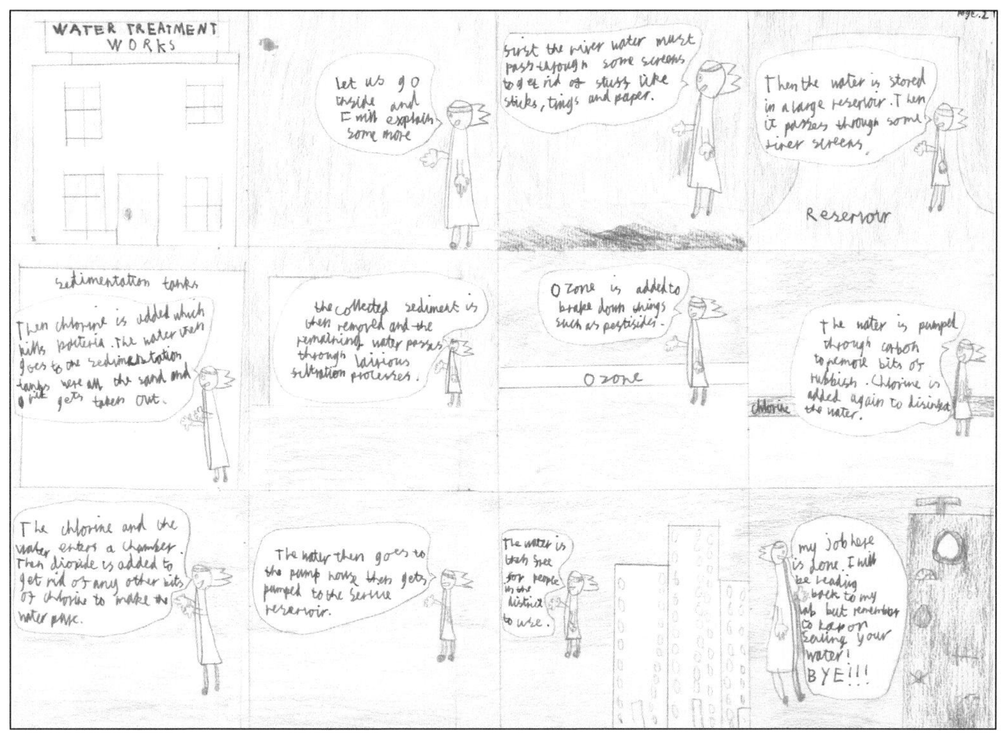

Stephanie Martin, aged 10 (Three Valleys Water).

Thomas Mattin, aged 10 (Three Valleys Water).

A2 River Road
Dhaka
Bangladesh

Dear Three Valleys Water

I am being helped to write this letter by my Teacher. I would like to ask if it is possible for you to come to my village in Bangladesh and help us install a water supply. I would like to tell you about myself and help you understand our problems.
I live near Dhaka with my Mother, Father two brothers baby sister and my Aunt and Uncle. Every morning at 5am I must get up and walk to the river Meghna with three buckets to fetch the water for my family. I help my family each day before school by breaking stones for a new road which my Father says may bring lorries with clean water one day.
My mother makes carpet from the jute grown in the countryside and my Father and Uncle sells them in the market each day. My sister is always ill and my mother tells me it is because the water is bad.
Each day in June for three weeks the floods come and our water becomes unsafe to drink. We have lost our home three times because the river becomes too full.
My teacher has lived in England and tells me that you make water safe for people to drink, wash and cook. I know that you are many kilometres away from our village but I have seen a picture in my school book showing how Three Valleys Water get clean water to many villages in Hertfordshire England. Our class is drawing pictures of how clean water willl help our families.
Please write and tell us if you can help us.

From Ahmed Bafhran

Written By Thomas Mattin aged 10

CHAPTER 6 Why is it so Important to Drink Water?

Research carried out in England has shown that drinking plenty of water during the day helps us in several important ways:

- It helps us to concentrate for longer periods
- It makes us less tired
- It makes us more alert
- It helps to prevent headaches.

Children should drink plenty of water every day. The amount you need depends on:

- Your body weight
- How active you are
- What the weather is like where you are.

If you don't drink enough water you could become dehydrated. This means the tiny cells that make up your body don't contain enough water to work properly.

Your brain is made up of millions of brain cells that need a lot of water. If they don't get enough water you don't think as clearly, you can't concentrate properly, and you may get a headache.

Thomas Moore, aged 10 (Severn Trent Water).
Thomas says: 'I drew a picture of my water bottle from Trent water which I fill up from my tap every day. The picture inside the bottle shows the mountains where rain comes from.'

Did you know?

- The average human body contains about 40 litres of water.
- We must keep an even balance of water in our bodies if we are to stay healthy and work well, so any water that is lost from our body each day must be replaced.
- It is better to drink water rather than squashes or fizzy drinks. Water doesn't rot your teeth, or upset your body by giving it too much sugar and other additives.
- You can survive for a few weeks without any food but without water you will die in a few days – take my word for it and do not try it at home or anywhere else!

Water enables valuable nutrients to be used by the body. It is involved in the transport of all the vitamins, minerals and other essential things around the body through the blood stream in a watery substance called plasma (which contains about three litres of water). Waste material is also carried in the plasma. This process is called the circulatory system.

The Promotion of tap water by water companies in England and Wales

As you now know, we should all have plenty of drinks every day and water is one of the best you can choose. The water companies in England and Wales all want to encourage everyone to drink much more tap water. They often publish information about water and the water cycle on their websites and provide games and books about water to schools. Many water companies have given sports-capped bottles to schools to encourage the children and teachers to drink more water.

The Tap Idols

One water company came up with a brilliant idea to encourage more people to drink tap water, especially those who prefer to have their drinks out of a bottle.

They based their idea on the TV show 'Pop Idol', only they used the name 'Tap Idol'. The search began to recruit four 'Tap Idols' as part of the 'Fill and Carry' campaign. This campaign was designed to encourage people to fill plastic bottles with tap water to take with them when they went out.

The company held auditions like those on 'Pop Idol' where the contestants had just two minutes to impress the judges with their personality, enthusiasm and ability to get the 'drink tap water' message across. Just imagine how difficult it must have been – just like performing in front of the judges on 'Pop Idol'! How nerve wracking!

The auditions were shown on the local television's news programme and the final four, two boys and two girls, were chosen to go out and about in the region and promote the scheme. The Tap Idols, who were also called 'Thirst-aiders', handed out 30,000 special re-usable bottles in twelve towns and cities across the region.

Do you think 'Tap Idols' is a good idea?

Would you be more likely to drink tap water if you had met these four while you were out in town with friends or family?

What would you do to get more people to drink tap water?

Ask your teacher if you could act out your own auditions using volunteer Tap Idol contestants – there is usually someone who wants to be a star!

CHAPTER 7 Questions and Answers

Here are all the questions that appear throughout the book – but this time with their answers.

Page 2
Question: Which do you think is lighter, water or ice?
Answer: Water as ice is lighter than water as a liquid. Ice floats.

Page 4
Question: What sort of temperature do you think is required for greater evaporation to take place?
Answer: High Temperatures. The sun is the main source of heat and as there is a lot of heat in hot climates this will cause water to evaporate faster.

Page 5
Question: What are clouds made of?
Answer: Clouds are made of tiny droplets of water (or ice or snow in very cold temperatures).

Faye Todd, aged 10 (Northumbrian Water).

Page 6
Question: Does the water cycle ever stop?
Answer: No

Page 6
Question: What do you think would happen if the water cycle ever stopped?
Answer: All living things would eventually die as our water sources would dry up. There would be no farming and so no food. Water is essential for life and without it we would not survive for very long.

Page 8
Question: Can you name the world's oceans?
Answer: Pacific, Atlantic, Indian, Southern and Arctic Oceans.

Page 9
Question: We have talked about droughts happening when there is not enough rainfall. What do you think would cause a flood?
Answer: Too much rainfall can cause rivers and streams to overflow and cause flooding where people live. Floods may also occur if drainage in an area is not very good. The water will not be able to run off into streams and rivers, or soak into rocks and soil.

Ellen Hornsby, aged 10 (Tendring Hundred).

Page 13
Question: Do you know what AD means and what language it is from?
Answer: Anno Domini. The language is Latin and in English it means 'the year of our Lord'.

Page 18
Question: What caused the water from the Thames River to be called 'monster soup'?
Answer: It was very dirty and smelly because of the sewage and waste that were thrown into it.

Page 20
Question: Do you think that we have benefited from inventions made by ancient civilisations? Can you name any of these that are in use today?
Answer: These include the use of alum, water filtration, irrigation and canals.

Page 21
Question: Water that is stored in rivers and streams is called surface water. What is water that is stored in underground rocks called?
Answer: Groundwater.

Page 22
Question: Can you remember what the rocks that store water are called?
Answer: Aquifers.

How did you do?

How well did you do in answering the questions?

11-12 Superb What can we say? Few rivers run as deep as your knowledge.

9-10 Excellent What a fountain of knowledge you are. You probably know more about water than your teacher!

7- 8 Very good You are well on your way to becoming a water expert.

5-6 Good With more practice you could turn your stream of correct answers into a huge river.

3-4 Not half bad You now have a pool of knowledge. Keep working hard and your pool will soon turn into a lake.

0-2 Not bad but watch out! You are starting to sink into the sea of 'don't know'. With some help you can do much better.

Glossary

AD: This stands for 'Anno Domini' and is the Latin for 'in the year of our Lord'. It is often used with a date to tell us that the year is after Christ was born.

Atmosphere: A layer of gas that surrounds the earth.

Aqueduct: A construction for transporting water over a long distance, often carrying it across valleys.

Aquifer: A layer of rock underground that can hold water. These rocks contain tiny holes called pores.

Archimedes' Screw: A device for raising water from one level to another. Consists of a giant screw inside a cylinder. One end of the cylinder is placed in the water and the other end is placed at the higher level. When the screw is turned, water is drawn up.

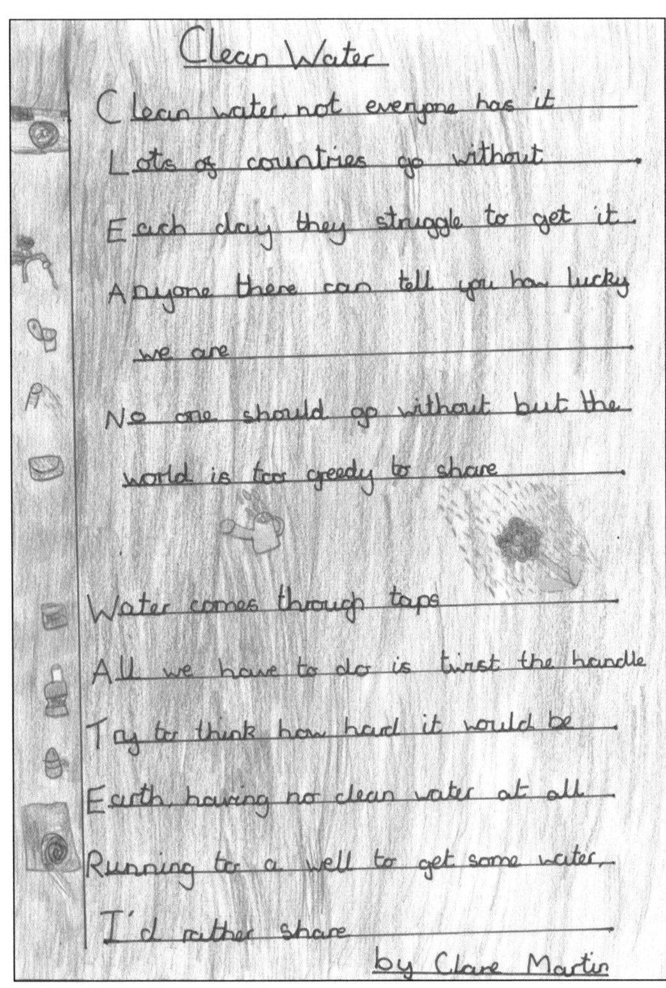

Clare Martin, aged 10 (Essex and Suffolk Water).

Atom: A very, very small particle, even smaller than a molecule. Atoms are so tiny that they cannot be seen through the most powerful microscope. A speck of dust is made up of a million billion atoms.

Bacteria: A group of very tiny organisms called microbes. 'Microbe' is a name given to all creatures that are too small to be seen without using a microscope. Some bacteria can cause illnesses, but not all bacteria are harmful.

BC: Before Christ. A term that we use with a date to show that the year was before Christ was born.

Canal: A man-made waterway.

Clouds: Collection of tiny droplets of condensed water.

Condensation: The process where a gas (such as water vapour) changes into a liquid (such as water).

Dehydration: The removal or loss of water.

Disinfection: A way of getting rid of harmful germs with the use of chemicals.

Drought: A long spell of dry weather when there is no rainfall, which will result in little or no water being available.

Environment: The surroundings and conditions in which people, animals and plants exist

Evaporation: The process where a liquid (such as water) changes to a gas (such as water vapour)

Flood: A large amount of water that collects on land, and is unable to drain away.

Fog: Mass of condensed water, like a cloud, that is near to the ground.

Gas: An airlike substance that is neither liquid nor solid.

Glaciers: Very slow moving masses of ice that have formed by the accumulation of snow.

Groundwater: Water stored in rocks underground.

Hydrology: The study of water.

Irrigation: The supply of water to land for the growing of crops

Lake: A large body of water.

Molecule: This is a very small particle but it is not as small as an atom. It is normally made up of at least two atoms.

Nutrient: Something that provides us with food.

Plasma: The liquid part of the blood. It contains about three litres of water (in an average human).

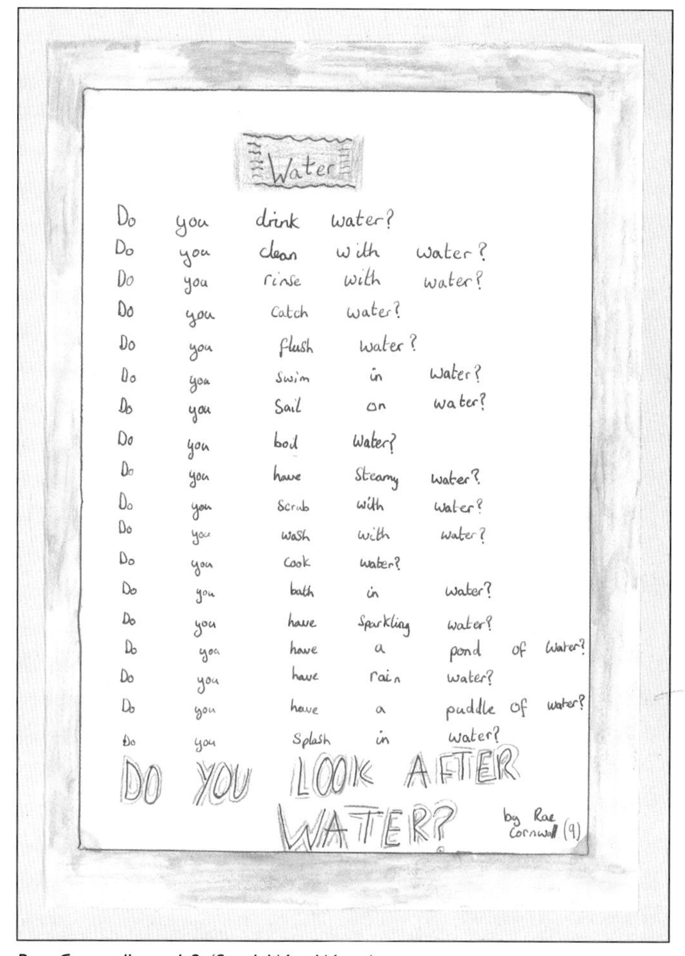

Rae Cornwall, aged 9 (South West Water).

Pollution: Materials and waste products from factories, humans, animals etc that are harmful to the environment, and that dirty or contaminate water, air, or land.

Population: All the people who live in a particular area, such as a village, town, city, county or country.

Precipitation: The process by which water falls from the atmosphere to the surface of the earth as rain, hail, sleet or snow.

Reservoir: A natural or artificially-created lake used for storing water. Reservoirs can also be underground.

Roman Empire: The area controlled by the Romans, stretching across Europe and beyond. The Roman Empire lasted from 509 BC to about 410 AD. The Romans settled in Britain around 54 AD.

Surface Water: Water that is found on the surface of the earth in streams, rivers, lakes and oceans.

Temperature: A measure of how hot or cold something is.

Transpiration: The process by which water is evaporated from plants. The plants give off water through their leaves.

Waterborne: Something that is carried in or by water.

Water cycle: The circulation of water from the ocean to the atmosphere, to the land and back to the ocean. This cycle is continuous.

Water Source: River, stream, lake, ocean, reservoir, aquifer, well.

Water vapour: The gaseous form of water.

How to build a water clock

The ancient Egyptians were very interested in time and invented many clocks. The water clock, or Clepsydra, was one of them. Clepsydra means 'water thief' and comes from the Greek words klepto meaning thief and hydro meaning water.

TO BUILD A WATER CLOCK YOU WILL NEED:

- Two disposable cups of the same size
- Paints (optional)
- A pen (that can be used to write on plastic)
- A piece of wood – about 20 cm long
- Glue

- Decorate both cups if you wish to do so. You could paint some Egyptian Hieroglyphics on the cups. You may already know this, but hieroglyphics are a form of writing using picture symbols. They were used by the ancient Egyptians.

- Make a small hole in the middle of the bottom of one of the cups using a sharp instrument such as compass point. Perhaps you should get a parent or teacher to help you.

- Glue one cup at the top of the piece of wood. Place the wood upright against a wall, making sure it is secure. Put the second cup under the first. Cover the hole in the top cup with a piece of tape and fill with water. Carefully remove the tape and start timing the water as it drips through into the second cup. Using the pen, mark the level of the water in the inside of the top cup at one-minute intervals.

- When the top cup is empty, pour the water from the bottom cup back into the top cup and you are now ready to tell the time by watching the water level as it drops down past your numbers.

- *You can see drawing of an Egyptian water clock on page 14.*

Wordsearch

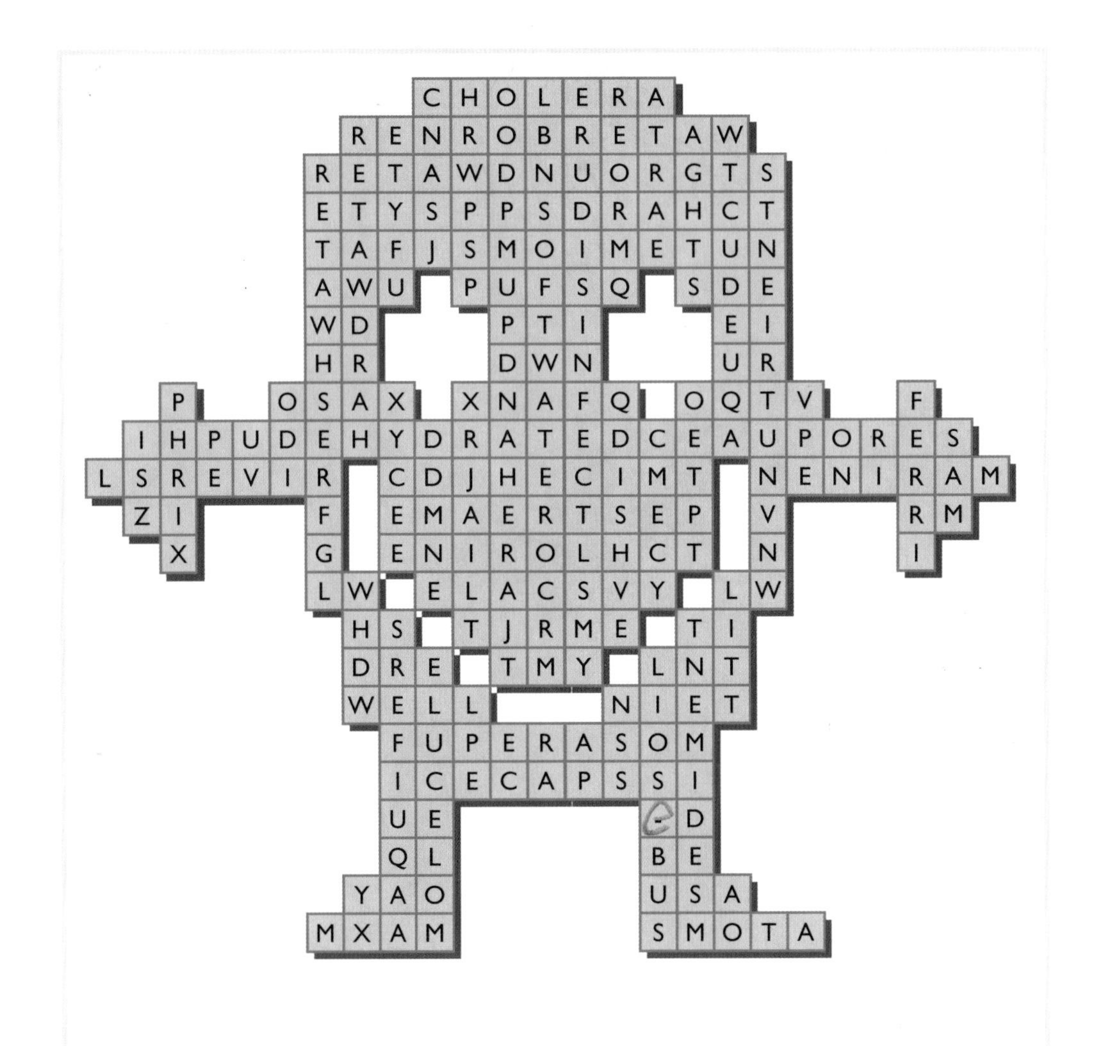

Word List

Marine	Soft Water	Chlorine	Hard Water	Pores
Reservoir	Stream	Dehydrated	Disinfect	Well
Scale	Waterborne	Aqueduct	Nutrients	Groundwater
Sediment	Aquifers	Atoms	River	Ice caps
Handpump	Cholera	Fresh Water	Sub-soil	Molecule